HABITATS

SAVING WILD PLACES

HABITATS

SAVING WILD PLACES

Dorothy Hinshaw Patent

ENSLOW PUBLISHERS, INC.

Bloy St. & Ramsey Ave. P.O. Box 38
Box 777   Aldershot
Hillside, N.J. 07205 Hants GU12 6BP
U.S.A.   U.K.

The author wishes to thank Neil Andre, Barbara Earnest, Norman Gershenz, Bob Jones, Nancy Marx, and Craig Tufts for their help with this book.

Library of Congress Cataloging-In-Publication Data

Patent, Dorothy Hinshaw.
    Habitats: saving wild places/Dorothy Patent.
        p. cm.—(Better earth series)
    Includes bibliographical references and index.
    ISBN 0-89490-401-9
      1. Habitat conservation—North America—Juvenile literature.
    I. Title. II. Series.
    QH77.N56P38   1993
    333.95'16'0973—dc20   92-28082
                                             CIP
                                             AC

Printed in the United States of America

10 9 8 7 6 5 4 3 2 1

Illustration Credits:

Neil Andre, p. 55; William Muñoz, pp. 40, 46, 49, 75, 80, 87; Dorothy H. Patent, pp. 8, 17, 19, 21, 24, 26, 28, 35, 71, 83, 91; Ken Schles, pp. 59.

Cover Illustration: William Muñoz.

Contents

1
Animals Need
Homes

North America is an incredibly varied continent. The huge expanses of treeless northern tundra, where caribou and wolves roam, and the southern alligator swamps, where the dense trees and brush are broken only by stretches of open water, could be on different planets. The evergreen forests of the northern Rocky Mountains, the cactus deserts of Arizona, the grassy midwestern plains—all of these are part of our continent. Not only do such areas look different, but they are home to very different assortments of plants and animals.[1]

Just like people, plants and animals need homes. All living things have basic needs that must be met. They need water, nourishment, and a place to stay. They need the right kind of environment for reproducing. People are unusual in that they can make comfortable homes in the desert or forest, in hot places and cold ones, on the mountaintops

Preserves like The Nature Conservancy's Pine Butte provide vital habitats for plants and animals in today's crowded world.

and in the valleys. Plants and animals are different. Each kind, or species, of wild thing can only survive under certain conditions. A lizard that thrives in the desert would quickly die in the cold northland, and an arctic caribou wouldn't last long in the desert. Each species has its own requirements for survival—range of temperature, amount of moisture, kind of food, type of surroundings, and so forth. All of those conditions taken together make up the habitat, or home, in which the species can live in a healthy way and can reproduce its own kind, insuring the survival of the species through time.

There are many different habitats, such as a mountain stream and its shores, a cactus desert, a grassland, or a rocky coastline. Each habitat features its own collection of plants and animals that share it and that depend on one another for survival.

Disappearing Homes

Unfortunately, homes for wild things are disappearing at a very fast rate in our crowded modern world. The human population keeps growing at a rate of about 2 percent each year. That means there are around 100,000,000 new people each year who need food, shelter, and water. Trees are cut down to build houses and to open up land for growing crops. Rivers are dammed to provide electricity, irrigation and drinking water. Marshes are filled to make land for shopping centers and new homes. Every time a piece of wild land is converted into human use, some plants and animals lose their habitat. Loss of habitat is the main reason so many species in the world have become endangered—and it's up to us to help save their habitats for them.

We hear a lot about the destruction of the tropical rain forests, especially the ones in South America. If the cutting and burning of the rain forests continues at its present rate, 80 percent of them could be gone by the year 2000. The rain forests are home to a tremendous variety of living things—tens of thousands of kinds of plants, insects, birds, amphibians, reptiles, and mammals. When the forests are cut down, the plants and animals die because their habitat has been destroyed. More and more species disappear. The forests are cleared mostly to create farmland for settlers. But the rain forest soil is not very rich, so after a few years, it is no longer good for growing crops. Then only grass will grow, and cattle ranchers move in with their herds. Even the grass doesn't last for long; soon the land is completely worn out. People and their animals move on, and the ruined land does not recover.

We do not hear as much about habitat destruction in our own country, but it is happening here, too. Whenever forests are logged by clear-cutting, animals, such as the birds that nested in those trees, lose their homes. When a swamp is drained to build a shopping center, the cattails, water lilies, frogs, and fish die. The damming of a river to create a reservoir destroys the homes of all the animals that lived in the valley that becomes flooded. It also alters the river in ways that can harm populations of fish and water birds.

How Habitat Loss Leads to Endangerment

There are a number of ways that habitat loss can endanger a species. The spotted owl, for example, can only survive in the moist, old-growth forests of the northwestern coast of North America, from Oregon through British Columbia in Canada. It builds its nests in old, dead trees. Each pair of owls needs quite a large area of forest in order to find enough food for themselves and their young and to be protected from predators. Trees that are 250 or more years old dominate these forests. Logging companies like logging the old forests, because each big tree can be turned into a lot of lumber. One tree may be worth several thousand dollars. By the end of 1990, about 90 percent of these forests in western Oregon and Washington had already been cut down.[2] Since the spotted owl has been declared an endangered species, the amount of the remaining old-growth forest that can be harvested is now being limited.

Some animals require habitats scattered over thousands of miles for survival. These are the migrators, animals that spend part of the year in one area and part in another. Monarch butterflies that live in the eastern and midwestern United States spend the winter in very concentrated areas of the Mexican forest. Fortunately, these winter sanctuaries are being preserved. Without them, it wouldn't matter how much space there was for the monarchs in the United States. Without a safe place to pass the winter, the species would die out.

Many of our familiar backyard birds also fly south for the winter. These birds also need their winter homes as well as their summer ones preserved. So much forest has been cut down in the Caribbean and Central and South America that the number of birds like wood thrushes and scarlet tanagers nesting in the United States has dropped significantly during the 1980s.[3] Habitat destruction in America hasn't helped either. Development that destroys forests also breaks the remaining woods into small sections, making it easier for predators to get at the nestlings during the summer breeding season.

Other birds, such as certain kinds of ducks, are having problems because nesting areas have become scarce. Ducks depend on wetlands like marshes and upland areas near the water for nesting. But wetlands are being drained for cropland, housing developments, and shopping malls. As wetlands decrease, there are fewer and fewer areas the ducks can find with the habitat they need. In some parts of the northern United States and Canada, where many ducks nest, the wildlife has become more and more concentrated into small areas. That means that good nesting sites are hard to find. It also means that predators like foxes are more likely to find duck nests and eat the eggs. In some areas, as many as 90 percent of duck eggs are eaten by predators.[4]

Ways of Preserving Habitats

Fortunately, many people are concerned about the loss of habitats around the world. They realize that we must share the earth with other living things. They know that we depend on the rest of the living world for our own survival. They also understand that humans have a need for nature in their own lives. People like to be able to retreat into quiet green places and listen to the sound of a stream rushing by. We enjoy watching birds and other wild animals go about their business. We can regain a feeling of balance in our lives by visiting a park and hiking away from human habitations.

Wildlife habitat is protected in a variety of ways. Governments preserve land as parks and refuges. Private conservation organizations buy important areas of wildlife habitat to preserve it. Citizens in many communities also make the effort to set aside land for wild things. Governments sometimes require developers to leave some land in its natural state when they build subdivisions. And individual citizens help by turning parts of their yards into wildlife habitats. By working together on all these levels, we can help insure that wild places are preserved.

2
A Freshwater Marsh

Across North America, there are many different types of habitats. A freshwater marsh is an especially rich habitat, so it makes a good example to show how the living things in a habitat interact with their environment and with one another. An enormous variety of life, from microscopic bacteria to huge moose (in northern regions) may make their homes in the marsh. The key ingredient of a marsh is an abundance of water. The soil is always wet, even if there is no standing water. Some marshes are merely meadows with ground that is constantly soaked. Others have stretches of open water from a few inches to several feet deep. But in all marshes, there are few, if any, trees. A water-soaked area with trees is called a swamp. Many of the same plants and animals live in both marshes and swamps.

If we could watch a marsh over a period of many years, we would see it change over time. Marshes are often transitional areas, representing shallow lakes and ponds in the process of filling up. Each year, the dead leaves of marsh plants add a new layer to the bottom, making the water a bit shallower. If streams feed into the marsh, silt brought by them can also contribute to filling in the marsh. Marshes, however, can last a long time. Some of the prairie pothole marshes in the Great Plains have been around for 11,000 years.[1]

In any habitat, plants are what make all the variety of life possible. Green plants contain a chemical called chlorophyll that allows them to trap the energy in sunlight and use it to manufacture sugars. The energy in those sugars is used by the plants for growth and for stored energy. When animals eat plants, they are acquiring the plants' energy for their own uses.

Marshes have different types of plants. Cattails and bulrushes are examples of plants that are rooted in the bottom ooze and send their tall stems up through the water into the airy sunlight. Many marsh plants are floaters. The saucerlike leaves and giant flowers of water lilies adorn the surface, connected to their underwater roots by long, flexible stems. Tiny bright green duckweed plants float delicately on the surface with the breeze. Each little duckweed plant has a thin root that dangles in the water below the leaf. Other plants live completely under the water, although some of them send up flowers to bloom in the

Freshwater marshes are especially rich habitats.

air. A variety of simple plants, called algae, also live in marshes.

A great assortment of animals also inhabit marshes. Tiny worms live on the bottom, feeding on dead leaves. If you look closely into the water, you may see many kinds of small swimming creatures, such as seed-shaped crustaceans, called zooplankton. Many animals lay their eggs in the quiet marsh waters. Their offspring, such as tadpoles, water dragons (dragonfly larvae), and mosquito wrigglers, are easy to find. Water beetles and even spiders join the diverse marsh community.

Marshes are home to many kinds of larger animals, too. Red-winged blackbirds sing from the tops of the cattails while American bitterns stalk quietly through the reeds. A variety of ducks nest in the protection of the dense vegetation while fish, bullfrogs, garter snakes, water snakes, painted turtles, and snapping turtles all live in and near the water. With so much prey to choose from, raccoons, foxes, and coyotes visit marshlands in search of food. Deer visit to drink, and in the north, moose wade out into the water to feed on the abundant plant life.

The plants and animals of the marsh are interdependent. The zooplankton feed on the algae suspended in the water. Small aquatic insects and fish eat the zooplankton. Meanwhile, tadpoles feed on clumps of algae and water plants. The tadpoles and small fish become prey for the water dragons, water beetles, spiders, and larger fish. At the same time, worms, crustaceans, snails, and the larvae of

Subalpine forests are a common habitat in the western mountains of North America.

other insects, such as stone flies and caddis flies, are feeding on decaying leaves on the pond bottom. These worms and insects become food for the fish. The fish in turn are eaten by top predators like herons, mink, and snakes.

The Importance of Marshes

Marshes are an especially critical category of wetlands—a term that covers swamps, bogs, and ponds as well as marshes—yet they have been disappearing at an alarming rate as they are dammed up, filled to make farmland, and bulldozed to provide space for houses, shopping malls, and parking lots. Wetlands serve a variety of important functions. Wetlands along river flood plains help slow floodwaters and absorb much of the excess water when flooding occurs. The microscopic bacteria and other scavengers in marshes and swamps help remove pollutants from the water, especially sewage. Wetland water seeps into the ground, recharging the groundwater from which many of our towns and cities get their water supplies.

But most of all, wetlands are productive of life. They are perhaps the most productive of all wildlands for several reasons. First of all, wetlands always have enough water. In many other habitats, limited water restrains biological reproduction. In addition, the openness of marshes to the sunlight gives the plants access to plenty of light, and the sunlight warms the shallow waters, speeding up the life processes of the living things that inhabit them. Wetlands are excellent recyclers of their own abundance. The bacteria, worms, and fungi

Housing developments have replaced many saltwater marsh habitats in the world.

that decompose dead plants and animals become food for water animals, keeping much of the energy within the system.

Wetlands produce many animals of interest to humans, such as fish humans like to catch and birds they enjoy hunting. In today's crowded world, wetlands provide sanctuaries for an increasing list of rare and endangered species. And they provide opportunities for human education, recreation, and appreciation of the abundance of nature.

Unfortunately wetlands are disappearing at an alarming rate in America. In the United States, 90 percent of our original wetlands have disappeared. Today, much of what remains is protected as part of the National Wildlife Refuge System or in parks such as Everglades National Park in Florida. Conservation groups around the country are working to protect more wetlands. But water is a big problem for wetland preservation. It is one thing to save a marsh. But if the river that feeds it dries up because it is used for crop irrigation or as a city water supply, the marsh will be destroyed. Groups like The Nature Conservancy are helping with this problem by buying water rights to rivers that feed refuges. These rights then are transferred to the refuge so that some water flow into the wetlands is guaranteed.

3
Public and
Private Lands

Setting aside large areas of land is especially important for preserving the variety of living things. Scientists have learned that the smaller the area, the lesser the number of different species that make their homes there. Each species that lives in a habitat needs to have enough to eat. Each needs certain conditions so that it can reproduce. A small area can meet the needs of some species but not others. And once a particular species disappears, others that depend on it in some way for their own survival may also disappear.

In the United States, an assortment of government and private organizations have set aside lands of different sizes. The amount of protection that living things in these areas receive varies enormously, depending on how the land is managed. Chapter 9 discusses the National Wildlife Refuge System, which is managed specifically for the benefit of wildlife by

Habitats can be preserved in a variety of ways. The marsh in the foreground was set aside when a housing development destroyed nearby marshland. The land around the church in the background was bought by the Landmark Society, in Tiburon, CA, to preserve its unique plants. The Landmark Society is working with other groups to preserve the hillside lands to the left of the church in the photo.

the U.S. Fish and Wildlife Service. But in addition to refuges, a variety of lands provide habitats for wild plants and animals in the United States.

National Forests

Our national forests are managed by the U.S. Forest Service, which is now 100 years old. National forests cover close to 163 million acres across the nation (this figure does not include the grasslands or research and experimental forests also managed as part of the system).[1] The forest must serve a variety of public interests, and that is a problem. They are important recreational areas, with campgrounds and hiking trails. They also contain large areas of wildlife habitat. But a third function of our national forests is providing timber for private companies to harvest. Once, forests were sometimes logged selectively. Only some of the trees were cut down. But now, because of the huge modern machines used for logging, almost all logging in our national forests is done by clear-cutting. Every tree is cut down, leaving a totally treeless landscape where plants and animals once lived in the shade of the forest.

National Parks

The National Park System began with the establishment of Yellowstone National Park in 1872. Yellowstone was the first national park in the world. The National Park system covers about 80 million acres and includes about fifty parks

Much of our national forest lands are logged, as can be seen from the clearcuts in the left of this photo.

and more than 250 other areas such as historic sites, battlefields, lakeshores, scenic trails. [2]

National parks are managed to allow people to enjoy nature in a variety of ways. They have campgrounds and often cabins where visitors can stay. Campfire programs in the evenings educate the public so they can enjoy and appreciate the park. Visitor centers feature displays explaining the natural phenomena within the park, and stores sell food, books, and equipment such as binoculars.

Park Service philosophy is to let nature work in its own way as much as possible within park boundaries. People can fish in the parks, but they can't hunt or cut down trees. The plants and animals are left alone as much as possible. Unfortunately, many parks were altered drastically before this philosophy was put into place. For example, wolves were hunted to extinction within Yellowstone National Park and all around it. Now, the deer, elk, and bison that wolves hunted are left alone, but without wolves to help control their populations. Many people believe that there are too many of these species in the park and that they are harming their habitat by eating too much of the vegetation.

Other Public Lands

States also have systems of parks and refuges. These lands are managed in a variety of ways, depending on the state. Generally these lands include campgrounds and picnic areas for people as well as homes for wildlife. Counties also may protect natural areas. In some places, such as the East

City parks provide an opportunity for people to enjoy the out-of-doors, but they provide habitats for only a limited variety of plants and animals.

Bay area of San Francisco Bay, the different public agencies that control public lands cooperate with one another in letting people know about the opportunities in the area for enjoying nature.

Cities also set aside land for parks. But generally city parks provide just one kind of habitat. They usually have broad, grassy areas with scattered, large trees. Such a habitat is great for picnics and for children to play. But it can be home for only a limited number of animal species, such as tree squirrels, worms, and songbirds such as robins. The grasses and trees are generally not species native to the area, and the habitat is maintained through artificial means such as watering, mowing, and using chemicals that kill plants other than the grass and trees.

The Nature Conservancy

In addition to government efforts to preserve habitats, private groups buy land to save it. Around the country, nonprofit organizations work to protect lands from development. The Nature Conservancy is the largest of such organizations and has worked since 1951 to identify and to protect important habitat areas in the United States and in other countries. Buying critical habitat is the main method the conservancy uses. The Nature Conservancy has brought protection to over 5.5 million acres of land in all fifty states and in Canada. It keeps the land and hires managers to take care of it when that method is the best way. As a result, it owns the largest network of private preserves in the world, with 1,200 preserves.[3] Sometimes, when it makes more sense to do so, the land is turned over to the government to be included in parks and refuges instead. The Nature Conservancy has also helped protect millions of acres in Latin America, working together with local preservation organizations.

Acquiring easements is another way to save habitats. An easement gives the rights to use land in certain ways to parties other than the owner. The most familiar type of easement is one that allows a power company or government agency to bury power lines or sewage lines on someone's property and lets them dig on the property if necessary to service the lines. A conservation easement is something different. Let's say a farmer lives near a growing city. As the city and its suburbs get closer to the farm, the value of the farmland increases and land becomes more desirable for subdividing. The more valuable land is, the higher the taxes the owner has to pay. Higher taxes make it harder and harder for the farmer to make a living and increases the temptation to sell the land for housing developments.

Farms often contain valuable wildlife habitats. A pond used to collect irrigation water, a farmer's wood lot, grassy pastures where cattle and horses graze—all these provide homes for wildlife. A farmer who wants to stay in business can avoid higher taxes by giving away the development rights for his land to an organization like The Nature Conservancy. This is called a conservation easement. The conservation easement makes it illegal for the land to be developed, which in turn makes the land less valuable, thus decreasing the taxes on it. The farmer can keep farming and paying lower taxes, but he and his heirs cannot turn the property into a housing development or sell the land for development.

4
In Your Own Life

There are many things each of us can do in our everyday life to help protect wild habitats. Everything in the world is connected with everything else, a fact most people have not realized until recently. What you do in your home and how you live your life contribute to the health or disorder of our planet. One person's actions may not seem like much. But when many people become willing to change their lives in simple ways, the effects can be enormous.

Resources and Daily Life

Americans are quick to criticize native peoples of other countries when they chop down rain forests to create the farmland they need to feed their families. By destroying the forests, people are harming the world's natural resources, and that affects all of us. But most of us are unaware that Americans use more of the earth's resources than the people of any other country. The United States has only 5 percent of the world's population, yet we use 25 percent of its resources! We also produce 25 percent of all the world's waste.[1] There are other ways to measure the effects of a society on the environment. If we take commercial energy use as a measure, each American causes seventy times as much damage to the environment as a citizen of the African country of Uganda or the Asian nation of Laos. We even consume twice as much energy per person as do residents of the wealthy countries of Sweden, France, Japan, and Australia.[2]

Resources and Habitat

But what do resources have to do with habitat? The answer is, a lot. We get electricity from a variety of sources, most of which affect habitats in different ways. In the Pacific Northwest, much of the energy is obtained from hydroelectric power plants. Dams are built on rivers to back up water. Then the mechanical energy of the flowing water is converted into electricity. When a dam is built, part of a river and the surrounding land is destroyed. Where shallow water once flowed, deep water now stands. The woodlands and grasslands near the river are flooded. Large amounts of habitat are destroyed.

In addition, dams create impossible obstacles for fish like salmon that swim up the rivers from the sea to lay their eggs. Some dams have "fish ladders" built in. A fish ladder is like a set of stairs with water flowing down it. The fish

Grand Coulee Dam in Washington state keeps salmon from swimming up the Columbia River to their spawning streams. As a result, the above-dam salmon populations are extinct.

jump and swim their way up the ladder to get past the dam. At least some fish are able to get to their breeding streams. But other dams are too tall for fish ladders. Once such dams go into operation, the salmon populations decline. Animals that count on salmon for food also may not survive.

Coal and oil are other sources of energy for making electricity. Burning coal or oil produces heat used to make the steam that drives generators to make electricity. Using coal causes habitat destruction in two different ways. First of all, coal mining lays waste to huge areas of the landscape. In some states, the land must be reclaimed—returned to its natural state—after the mining is finished. But it is almost impossible to recreate the natural landscape after mining.

Burning coal to produce power is also a major cause of acid rain. Chemicals such as sulfur and nitrogen oxides are released into the atmosphere by coal-burning plants. The chemicals are altered in the atmosphere and return to earth as acid rain or snow, often hundreds of miles away from where they originated. The rain can be so acidic that it destroys forests. Acid rain can make lakes and streams so acidic that fish and insects cannot live there anymore. Sixty-five percent of the sulfur dioxide (the main ingredient in acid rain) released into the atmosphere in the United States comes from electrical utilities.[3]

Oil is used not only to generate electricity but is also refined into the gasoline that powers our cars and trucks. The emissions from such vehicles are a major cause of air pollution—20 percent of the carbon dioxide emitted into the atmosphere in the United States comes from cars and light trucks. In addition, private vehicles produce 34 percent of the nitrogen oxide, another major cause of acid rain, that is added to the atmosphere in our country.[4]

The more oil needed to generate electricity and to power automobiles, the more oil will be drilled from the ground. Oil drilling imperils habitats in more ways than one. The disturbance to the environment where drilling occurs destroys habitats. In addition, oil spills kill countless plants and animals. We have all seen tragic photos of blackened, dying sea birds covered with oil. But we do not see all the plants and animals under the water's surface that have also died from the oil. The long-term effects of oil spills on habitats are unknown.

Changing Your Life

All of these problems may seem beyond our control. But that is not completely true. We need electricity. Most Americans rely on their pollution-producing automobiles and pickup trucks for transportation. If each of us does our best to cut our energy and automobile use, there will be less demand for energy. Less demand for energy means less pressure to destroy remaining habitats to produce energy.

Simple changes in your life help. When you want to read during the daytime, sit by a window and use natural light instead of turning on a light. When you leave a room, turn off the television, stereo, and lights. Turn on only the lights you need instead of lighting up the whole room.

When the weather is good, ride your bike or walk instead of asking your parents to drive you places. When you do need to be driven, try to set up a car pool with others who are involved in the same activity. That way, only one car is driven instead of several.

Recycling also helps habitats in a variety of ways. Producing aluminum cans from mined aluminum, for example, takes more energy than making them from old cans. And mining destroys habitats. By recycling your cans, you are reducing the amount of energy needed and thereby helping the environment. In the same way, new paper and cardboard use up both trees and energy; recycling gives the environment a chance.

Recycling helps in another way, too. In our throwaway society, garbage takes up an enormous amount of space. Every acre of new landfill is one less acre of habitat for wild things. When we recycle, we decrease the amount of garbage that needs to be disposed of.

Creating grasslands to feed cattle is a major cause of rain forest destruction in Central and South America. The meat from these cattle goes largely to make inexpensive hamburger that is bought by fast food chains in the United States. Because of pressure from their customers, some fast food companies claim that they no longer buy this beef. In general, producing meat takes more from the environment than growing grain and vegetables. You can help the environment and do your health a favor by cutting down on the amount of meat you eat.

Groups Can Help

Perhaps you belong to an organization like the Girl Scouts (including Cadette Girl Scouts and Senior Girl Scouts), Boy Scouts (including Varsity Scouts and Venture Scouts), or Campfire Boys and Girls (including Discovery and Horizon clubs). These groups have a variety of projects that can involve habitat preservation. For example, Girl Scouts can earn an Earth Matters patch by participating in activities suggested in the book *Earth Matters: A Challenge for Environmental Action*, part of the Girl Scout Contemporary Issues Series. Girl Scouts also can earn an Eco-Action badge. Boy Scouts have a Scouting Environment Day in April for conservation activities and a program in conservation. Boy Scouts who are active in conservation may qualify for the William T. Hornaday awards, which recognize distinguished service in conservation.

Clubs like Kids for Saving Earth exist especially for conservation purposes. Kids for Saving Earth was started in 1989 by 11-year-old Clinton Hill at his school in New Hope, Minnesota. Clinton died of a brain tumor before his twelfth birthday, but his club lives on. By mid-1991, more than 5,000 Kids for Saving Earth clubs across the United States and in six other countries worked to make the world a better place for all living things. The organization provides information on hazards to the earth such as air pollution and habitat loss and tells how people can help change things. The word "kids" in the name of this organization may make you think this is a group only for young

A local Boy Scout troup helped build this artificial island for nesting Canada geese.

children—not so. Many chapters are in the middle and junior high schools. Some high school conservation clubs use the information from Kids for Saving Earth to help plan their activities.[5] If you are interested in joining this club, or would like to involve your class at school in a project, look in your phone book for the chapter nearest you.

Milestones in Habitat Preservation

1872—Yellowstone National Park, the first national park in the world, was established.

1903—Pelican Island, the first National Wildlife Refuge, was set aside.

1934—The Migratory Bird Conservation Act (Duck Stamp Program) was signed into law.

1937—The Federal Aid in Wildlife Restoration Act (Pittman-Robertson Act) was signed into law.

1951—The Nature Conservancy was founded.

1970—The first Earth Day celebration was held.

1972—The Marine Mammal Protection Act was passed.

1973—The Endangered Species Act was passed.

1988—A strengthened version of the Endangered Species Act was passed.

1990—The twentieth Earth Day was celebrated with great fanfare.

1991—The Madrid Accord was signed to protect Antarctica for fifty years from mineral exploration, mining, and oil drilling.

5
In Your Own Backyard

You can help create a habitat for wild plants and animals near your own home. As more and more of our wildlands are turned into housing developments and malls, fewer and fewer suitable habitats are available for wild things. With the help of your family, you can do your part to reverse this trend by turning part of your yard into a natural habitat. As a bonus, you will be helping provide yourself with opportunities for wildlife watching.

For starters, head for your local library to research native plants that you can use in your yard. The library should also have guidebooks for identification of the animals, including mammals, birds, butterflies, and others, that might show up once you have provided what they need to call your yard home. If there is an arboretum in your area, you can visit it to see native plants that might interest you. Visit a local store that sells bird feeders to check out what is for sale. If the feeders are too expensive for you to buy, you can make one yourself in your school woodshop or at home from recycled materials. By studying the feeders in stores, you can get tips on how to design your own.

Apartment Wildlife

Even if you live in a city apartment, you can help provide vital food for butterflies and birds. These flying animals can easily get from place to place to find food and water, so they can take advantage of city feeders. Bird feeders not only provide the birds with food, they also give you an interesting show as birds come to feed. You can attract different kinds of birds depending on what sort of seeds you use. Many birds like sunflower seeds—cardinals, gold-finches, purple finches, nuthatches, and woodpeckers, for example.[1] Be careful of inexpensive birdseed mixes. They are likely to attract non-native birds such as pigeons and starlings that can take over and bully native species. Look for mixes containing plenty of small, beadlike red and white seeds. These are proso millet seeds, preferred by native species such as cardinals.

Hummingbird feeders, filled with plain sugar water, can bring these beautiful birds to your windowsill or balcony where you can watch them. Choose a red plastic feeder. Hummingbirds are strongly attracted to red, the natural color of many flowers with abundant nectar. Hang the feeder near a window so you can watch the birds feed.

If you go on vacation, take the feeder down until you return.

In addition, windowboxes filled with flowers can attract hummingbirds, honey bees, and butterflies. All these animals are fascinating to watch as they feed. And if you plant herbs like thyme or sage, your family can also enjoy fresh herbs in cooking. Herbs mixed in with flowers that bloom all summer, such as marigolds, sweet alyssum, or sweet William, provide a beautiful and useful source of food for wildlife.

Planning a Yard

If you are lucky enough to live in a house with a yard, even a small one, you can do more to help wild things survive. Remember that every animal has a variety of needs. It needs food, water, and shelter. It also needs a safe place to raise a family. It is not difficult to turn your yard into a place where wild things can live and reproduce, if you and your family want to do it. Depending on how much time, effort, and money you can invest, you can make a few minor changes in your outdoor surroundings or completely redesign your yard to make it attractive to wildlife.

Although creating a wildlife habitat takes work at first, it actually decreases the amount of time needed to care for the yard once it is established. Most of our yards are occupied by lawns. Lawns across the country take up an area equal to the size of the state of Michigan![2] Native plants are adapted to the amount of rainfall in your area,

Squirrels are among the animals you can attract to your yard by providing food, shelter, and water.

so they do not need to be watered as often as does a lawn. If you replace part or all of your lawn with native plants, your family won't have to spend as much time and money watering, mowing, weeding, and fertilizing it. Many areas have nurseries that specialize in native plants. Check your phone book, or read advertisements in the newspaper during the early spring to find these nurseries, or call your local county agricultural extension agent to get the information.

Some examples of native bushes common in much of the United States that will attract wildlife are American elderberry, buttonbush, and silky dogwood. Elderberry bushes are very attractive, with lovely leaf fans and clusters of white flowers in spring. In the fall, the flowers develop into black berries that attract birds. The pleasantly scented flowers of buttonbush attract both birds and butterflies, and the shrubs also provide cover and food for wildlife. Silky dogwood is attractive at all seasons and can provide vital cover for wildlife. The purple or reddish twigs accent a winter landscape, and the creamy-white flowers bring butterflies in springtime. Birds enjoy the blue berries during the fall, and people appreciate the bright red fall foliage.

Buying plants can cost a lot of money, and planting a yard can take a lot of time. Even if you can't afford the time and money to redesign your yard, you can do simple things to make life easier for the wild animals that live nearby and to attract them to your yard. Like a city dweller, you can set up bird feeders. Chances are, your feeders will attract a variety of birds that change with the seasons.

Providing Water

Perhaps the easiest and most ignored way to help wildlife is to provide them with a source of water. Maybe you have watched birds gather to drink and bathe at puddles that form when your yard is watered. If you have a source of water they can count on at all times, you will be able to see the birds more often, and you will give them something they need badly. Other wildlife, such as butterflies, squirrels, and raccoons, may also come for a drink.

Any kind of wide, shallow dish can be used as a wildlife water supply. A few dollars will buy a large clay dish normally placed under a flowerpot to catch excess water. You can set out a number of dishes in your yard. One dish nestled next to a bush may attract chipmunks or toads, while one out in the sunshine will bring in birds.

Every few days be sure to check your containers, clean them out with a hose or small brush, and refill them with fresh water. If you live where winters are cold, providing water in the wintertime is more difficult, but it can be done.[3]

Providing Family Homes

Birds need protected places to make their nests and raise their families. By hanging a birdhouse in your yard, you

Birdhouses should be cleaned out after the birds leave to prepare them for the next season.

can provide a home for a bird family. You and your family can also have fun watching the family grow.

Each species of bird looks for certain requirements when it chooses a home. But a generalized birdhouse can attract a variety of species—chickadees, bluebirds, and swallows, for example. A house five inches wide and ten inches high, made of unpainted half-inch thick wood, will do. It should have three-eighth-inch drainage holes in the bottom and a one and one-half-inch entrance hole on one side near the top.[4] Perhaps you can make a birdhouse in a woodshop class at school, or you can buy one at a pet store or birdwatchers' store.

The house should go up in late winter or early spring, mounted on a pole, post, or tree, and should be between three and five feet above the ground. After the bird family leaves the birdhouse, you should clean it up for the next year. Some birds, like wrens, may return to the same birdhouse to raise their families.

Flowerbeds as Food and Shelter

You can attract more wildlife to your yard, especially hummingbirds and butterflies, just by carefully choosing the sorts of flowers in your flowerbeds. When your family is planning what flowers to plant, suggest some species that provide butterfly food and abundant nectar, and you can enjoy your garden more than ever at the same time as you help wild things.

One of the best plants to attract butterflies is milkweed. Milkweeds are perennials—that is, the plants die down in winter and come back year after year. So once they are planted, they can be enjoyed over a long period of time. From summer through early fall, milkweed produces lovely clusters of small flowers with abundant nectar and a delightful aroma. Many kinds of butterflies will visit milkweed blossoms. Milkweed provides an exciting bonus—it is the food plant for caterpillars of the beautiful monarch butterfly. If you are lucky, monarchs will lay eggs on your milkweed plants. Then you can follow the development of the caterpillars from tiny black-headed, grayish white hatchlings to spectacular black and yellow striped creatures with a pair of black feelers at each end of the body.

Milkweeds come in a variety of types, from common milkweed, with its pale pink flowers, to the red milkweed found along the Atlantic and Gulf coasts. The easiest species to get is butterfly weed, which is sold in stores. Butterfly weed is usually orange, but also comes in red and yellow. Once it is planted, it will come up year after year, providing your garden with beauty, and butterflies with food.

Annual flowers live only for one year and need to be replanted each spring. Many beautiful annuals, like impatiens, marigolds, phlox, and zinnias attract a variety of butterflies as well as adding color to the garden.

Generally speaking, plants whose flowers appear in flat clusters of small blossoms are most likely to attract butterflies, for they provide a platform where the insects can land while sucking nectar.[5] Hummingbirds are drawn to red flowers, especially ones like columbine that provide lots of sweet nectar. Like milkweed, columbine is attractive and easy to grow and comes back year after year.[6]

Making It Official

The National Wildlife Federation encourages people to turn their yards into wildlife habitats through its Backyard Wildlife Habitat Program. It offers a variety of materials to help plan a backyard habitat.[7] Once you've succeeded in providing food, water, cover, and places to raise young, you can apply to have your yard enrolled as an "official" wildlife habitat.

The National Institute for Urban Wildlife also has a program for certifying wildlife habitats, as do a few states.

6
Around Your
Neighborhood

Around the country, people have banded together when critical habitats have been threatened by development. They have filed petitions, helped pass laws, collected money, and purchased lands to keep wilderness around them. In Tiburon, California, for example, a group of citizens established the Landmark Society, which has preserved historic buildings and natural lands, some of which are habitats for plants that grow nowhere else.

Chances are that in your own neighborhood people have begun to band together to preserve or restore a wild habitat. It is happening in big cities as well as in suburban areas. Take a look around you. Is there a vacant lot where no building has stood for years? Is there a well-manicured park with no shrubs to provide cover and food for wild things? Does a company have extensive grounds around its buildings that are sterile of life? If so, see if you can help bring diversity and naturalness to such places. If there is not already a group devoted to bringing nature back into human surroundings, you can help get one started. You can help educate people to see that they can gain pleasure from green oases that harbor a variety of life, and you can lobby government and businesses to develop such havens.

The Dolphin Defenders

Even in the inner city, neighbors can do things to provide animals with homes. In Saint Louis, Missouri, a group of nine- to twelve-year-olds, who call themselves the Dolphin Defenders, have energetically attacked a variety of environmental problems, including litter and habitat preservation. The group is sponsored by Dignity House and helped by director Neil Andre. The Dolphins meet after school on Thursdays and Fridays. They often work on weekends on their more time-consuming projects. During the summer, they are together from 8 A.M. to 4 P.M. for the five working days in each week. Over a recent eighteen-month period, the thirty Defenders collected 112,000 aluminum cans. Recycling those cans amounted to an energy savings equal to 6,000 gallons of gas. Every penny the Dolphins raise through recycling efforts or other fund-raisers goes directly toward helping the environment. Adults can help by paying a donation, thereby becoming lifetime members of the Dolphin Defenders, or by buying voting stock in the Dolphins. Stockholders get to choose two of the many

Dolphin Defenders (clockwise, from upper left) Bryant Chambly, Demetrois "Ali" Roack, Clifton Jamieson, Tony Benson, Jason Harris, and Myron Love surround a young China girl bush on "The Promised Land," a wildlife habitat they helped establish.

activities the Dolphins carry out each year. All the other projects are chosen by the children.

The most enduring of the Defenders projects are their wildlife habitat establishment efforts. The group was founded in 1987. By the end of the summer of 1991, they had turned four vacant lots into habitat areas and had them certified as backyard habitats by the National Wildlife Federation. The fourth habitat was especially challenging. Called "The Promised Land" by the Dolphins, this habitat lies on a thirty-five-by-sixty-foot lot next to a church. The old buildings on the lot had been torn down, with much of the leftover rubble bulldozed into the ground. The church wanted to turn the lot into a parking area. But the Dolphins convinced the church to let them create a wildlife habitat instead and allow it to remain as a habitat for at least three years.

It was a tough job. The Dolphins removed over 800 pounds of exposed rocks from the lot. By the time they took away broken bricks and partially exposed rocks and bricks, they had hauled off over a ton of material. One eighty-pound rock took three adults working along with the children three hours to dig out and cart away. The remaining hole was just the right size for planting a rhododendron tree, using soil donated by the city parks system to fill the hole. Now that tree will provide cover for wildlife and beautiful flowers for people to enjoy. The Dolphins also planted small flowers like tulips and begonias and barberry bushes to provide food and cover for birds. They dug a hole and sank an eighteen-inch plastic dish for water to nourish wildlife, and built a brush pile in which animals can hide and make their homes.

Because the ground was in such poor condition, most of the plants were placed in raised beds. The Dolphins placed large concrete blocks together and filled the area inside with soil. There they planted evergreens and ground cover. They bored cavities in a dead tree that remained on the lot in hopes that birds would nest there, and they put up two birdhouses.

The Dolphins' efforts have paid off for wildlife and for their own enjoyment. On their habitats, they have seen a variety of birds, including blackbirds, chickadees, blue jays, rock doves, titmice, thrushes, and barn swallows. House sparrows, starlings, and cardinals have nested. Opossums, groundhogs, squirrels, rabbits, raccoons, and lots of butterflies have also shown up. One day, the lingering aroma of a skunk convinced them that this animal had at least paid a visit.

The idea of the Dolphin Defenders is spreading. So many children wanted to join that a new group of Dolphins that meets on Saturdays has been formed, and a club calling itself the Universal Unicorns has begun in another neighborhood.

The Dolphin Defenders have been commended by a variety of organizations, including the Humane Society of the United States, the National Audubon Society, and the Better World Society. They have testified in front of a government hearing attended by congressmen and a White House representative on how they achieve their goals. If the idea of the Dolphins spreads to other cities, more American wildlife can come to share the urban world inhabited by most of our people.[1]

The Green Guerillas

New York City is the ultimate city, with its skyscrapers so tall and dense that they block out the sun on much of Manhattan Island. Yet even there, people work to establish outposts of green. Vacant lots don't have to remain vacant in any city—they can be turned into useful and beautiful pieces of land. In 1973 the Green Guerillas began as a group promoting community gardening within the city. Executive Director Barbara Earnest explains, "In the beginning our main purpose was to show people how to start a garden. Our main purpose now is to preserve and improve existing sites."[2] The Green Guerillas provide free plants, help plan and tend gardens, identify pests, instruct on practical matters such as composting, and publish a newsletter. In 1990 alone, the Green Guerillas donated more than 9,000 hours of volunteer time to helping with gardens and collected more than $260,000 worth of plants to give away.

The gardens the Guerillas have helped establish provide welcome green relief from the hard, gray sidewalks and streets of the city. Some have benches where people can rest and view the garden and the birds, bees, butterflies,

It is hard to believe that the Liz Christie Bowery Houston Garden is in the heart of New York City, but work by energetic volunteers can help create such green areas for wildlife and people in any city.

and squirrels that visit. In addition to growing flowers and food for people, some of the gardens have also been developed as wildlife habitats. By 1989 two of the gardens had been registered as backyard habitats by the National Wildlife Federation. "Before the gardens, we had pigeons and not much else," says Sandi Andersen, who has a plot in El Jardín del Paraíso (the Garden of Paradise), on Manhattan's Lower East Side. Since the garden was established, Andersen has seen blue jays, cardinals, finches, woodpeckers, and other birds not normally associated with cities.[3]

The city followed the Green Guerillas' example in the late 1970s by establishing Operation Green Thumb. Today, this program provides leases for city-owned vacant lots for about 500 community gardens.[4]

7
At Your School

Schools provide plenty of opportunities for helping secure habitat for wild plants and animals. Students can choose an aspect of habitat conservation as a report topic. The more you learn, the more you will be able to help. Classes can take on projects to learn more about the needs of living things. They can raise money for preservation and donate it to an organization that preserves habitats. By working together with interested teachers and administrators, students can become important sources of help for the natural world. School efforts are getting underway all across the country. In Petaluma, California, children grow native trees for replanting in nearby habitats. In the Midwest, the Society for Ecological Restoration is showing teachers how to do ecological restoration work so they can knowledgeably involve their schools in such efforts.

Perhaps the most immediate way to help is to establish a nature area right on the school grounds. The students and teachers at Mead Junior High School in Spokane, Washington, did just that in 1990. An area on a hillside next to one wing of the school was officially dedicated at an event entitled "Saving the Present for Future Generations." Each homeroom that participated in the ceremony was given a Douglas fir tree to plant in the nature area. Science teachers planned ways to use the nature area as part of classroom studies—for insect collections, studies of edible plants, and growing of native plants. Shop students would build birdhouses to be set up there. The nature area would become a part of the school, just as much as the school buildings.

But a year after the nature area was dedicated, the school needed a place to put portable classrooms. The nature area appeared to some the best choice. Most of the staff and students were upset. After all, it was supposed to be a place that would continue to evolve and develop as a natural community. Certainly some other place could be found for the portable classrooms. Students made posters to promote saving the nature area. One of the science teachers wrote a letter to school board members pointing out the difficulty in teaching the students about how damaging the destruction of the rain forest and other habitats is when their own special habitat was in line for elimination. The students at Mead were learning firsthand how environmental issues so easily become secondary to

Did You Know That . . .

- Wetlands are home to about 170 species of plants and animals that are on the U.S. government's list of threatened and endangered species.

- Each U.S. citizen uses five times as much oil as the world average—that's about 1,000 gallons a year per American.

- Plastic debris is the most abundant litter on beaches—and accounted for 63.9 percent of the trash collected during the national beach cleanup in 1990.

- The United States has lost between 70 million and 100 million acres of open space since World War II.

- It takes only ten minutes for a logger with a chain saw to cut down a 1,000-year-old tree.

- Less than 5 percent of the world's tropical forests are protected within parks and preserves.

economic concerns. That is really what is behind all the destruction of the natural world—making or saving money is often more important to the people who run the world than other values, such as respecting life on earth, taking responsibility for the future of the planet, and understanding the human needs for contact with the natural world for our own well-being.

Fortunately for Mead Junior High School, another site was found for the portable classrooms, so the teachers and students can continue to enjoy and to learn from their nature area.[1]

Hard Workers in Florida

The students and staff of the Lee County schools in Florida are especially active in preserving the environment. They work under the leadership of William Hammond, director of Environmental Education and Instruction Development Services. Hammond was inspired by the first Earth Day in 1970. He realized that being aware of environmental needs was not enough—children wanted to help change things. He founded the Monday Group, a gathering of high school juniors and seniors. The Monday Group not only studies environmental problems but does something about them.

The Monday Group has accomplished a great deal, including conducting a campaign that resulted in saving a vital part of the Six-Mile Cypress Swamp preserve. The students met with citizens and county officials and planned television and newspaper publicity that helped convince voters to pay the $2.5 million needed to buy the property through increases in taxes. They were also influential in getting habitat set aside for two endangered species, the bald eagle and the manatee.[2]

The schools of Lee County have done more than help create preserves and parks. They are also transforming their school grounds into wildlife habitats. All fifty-five schools in the district are working on at least some landscaping using native plants. In addition, twenty-one schools have signed up for a special program for establishing habitats at the schools. Each year school representatives attend a day of training, listening to experts on such topics as soil, wildlife species, butterfly biology, and ponds. Parents, teachers, and children all belong to the development teams that work on establishing the habitats. The school habitats also provide opportunities for scientific research—at one school burrowing owls are being studied.

The schools cooperate with one another in their efforts. For example, the students of one school, which needs money to buy plants, will gather and bag pine needles. They sell the pine needles to another school to use as mulch—material placed on the soil surface to help keep down weeds and retain moisture.[3]

8
Restoring and Maintaining Habitats

It is very important to protect critical wildlife habitats that already exist. But backyard and neighborhood projects show how destroyed habitats can be renewed. Habitat renewal is important. As it is, wildlife has little habitat left in most parts of our country. We have cut down 90 percent of our temperate rain forests; only parts of the forests in the Pacific Northwest are left, and the timber industry is objecting loudly to saving any of that to protect endangered species like the spotted owl. The prairies, which used to cover hundreds of miles in all directions on the plains are almost totally gone; just a few small areas have been preserved. Our wetlands have also disappeared—both California and Connecticut have lost 90 percent, while 95 percent of Iowa's are gone.[1] Once a habitat has been destroyed, it is usually gone forever. But sometimes, if the land has not been paved over or built upon, it can be restored to something resembling its original condition. Restoration takes lots of time and money, but by restoring habitats, we can help compensate for their loss.

The Government and Restoration

The federal government has its own program for restoring wildlife habitats. A law passed in 1937, called the Pittman-Robertson Act, levies a tax on the purchase of firearms and ammunition used for sport hunting. The money raised is used in combination with state funds for a variety of projects, with most of it going to buy, develop, maintain, and operate wildlife management areas. Since the program started, about four million acres have been bought. That's as much as the area of Rhode Island and Connecticut combined. In addition, about forty million more acres are managed for wildlife, thanks to agreements with the land-owners.

Wildlife management areas provide critical habitats of many kinds. Big game animals like antelope, elk, deer, and bighorn sheep use different kinds of habitats through the seasons. Especially important for their survival is winter range where they can find enough to eat through the cold months. Winter range is generally at lower altitudes than summer range, and people like to live in the same kind of areas with milder climates that the animals need. Because of its popularity with humans, much winter range is privately owned. The Pittman-Robertson funds have enabled states to buy such land and to negotiate agreements with landowners to share their land with wildlife.

Pittman-Robertson money also goes to preserve and restore vital wetlands and to plant trees and shrubs on the Great Plains as protective cover for quail and other wildlife. In heavily wooded parts of the Northeast, trees are cut down to create clearings in the forests. The clearings provide habitats for deer, rabbits, grouse, and other wildlife. In the South, brush and tall grass are burned so that seed-producing plants will grow, providing food for wild turkey and quail.[2]

The Science of Restoration

A restored habitat is unlikely to be as rich as an original, undisturbed one, but it is better than wasted or barren land. In the San Diego area, people have tried to restore salt marshes. Restored marshes have fewer species, less rich soil, thinner cover, and fewer numbers of invertebrate animals than natural marsh. This is unfortunate because current law allows developers to destroy wetlands if they replace them with habitats they recreate somewhere else. It is always preferable to prevent the destruction of habitats in the first place. But in the real world, the new science of restoration biology gives some hope of providing much needed homes for living things.

In the Chicago area, Steve Packard of the Illinois Nature Conservancy has spent years working on restoring a vanished American habitat, the tall grass savanna.

Savannas are grasslands with scattered trees. Packard and his co-workers hunted in unlikely sounding places such as railroad rights-of-way and the edges of cemeteries for seeds in their hunt for the appropriate species of native grasses. They eventually succeeded, and now there are a number of restored savanna grassland areas in the Chicago vicinity.[3]

People around the country are becoming interested in the idea of habitat restoration, which has led to the foundation of the Society for Ecological Restoration. A journal, *Restoration & Management Notes,* is published twice a year to keep members and other interested parties informed about the problems and successes of the habitat restoration movement.

The Importance of Trees

Around the world, trees are being cut down at an alarming rate. In the Pacific Northwest region of our country, a battle rages between the logging industry and environmentalists about how fast the forests can be cut. In Central and South America, tropical rain forests are disappearing to make way for farms and cattle ranches. In Asia, as in the United States, forests are disappearing as trees are cut and sold to the timber industry.

Forests have value in a variety of important ways. For one thing, they are homes to countless numbers of species of plants and animals. When the forest are cut down, the plants and animals that rely on them disappear. But forests

Many animals make their homes in forests like this one.

are valuable for other vital reasons as well. About half of the greenhouse effect, which scientists fear will result in a deadly increase in the world's temperatures, is caused by an increase of the gas carbon dioxide in the atmosphere.[4] Carbon dioxide comes from a variety of sources, including the burning of fossil fuels to make electricity.

Trees are a great reservoir for locking away carbon. As a tree grows, it takes carbon dioxide out of the air and uses it as a building material to produce leaves, stems, and branches. Much of the carbon ends up stored in the wood of the tree, and oxygen is released into the air. In one year, the average tree can remove twenty-six pounds of carbon dioxide from the atmosphere! The burning of the rain forests also adds carbon dioxide to the atmosphere, increasing the greenhouse effect.

In addition to providing habitats and controlling the carbon dioxide content of the atmosphere, trees can perform other important tasks for our planet. When trees that lose their leaves in winter are planted on the south side of a house, they help cool the house in summer through their shade, reducing the need for energy-consuming air conditioning. In the winter, the sun can shine through the bare branches and help keep the house warm, cutting down on the need for heat. Trees planted in cities also help soak up pollution from the air, making it healthier to breathe.

Fortunately, there are organizations that encourage the planting of trees. Global ReLeaf, a program of the American Forestry Association, is working around the world toward this goal. It has established a network covering all 50 states and has launched an international program starting in Canada, the United Kingdom, Hungary, and Costa Rica. Global ReLeaf raises money for tree plantings all over the United States, and provides information to homeowners on how to plant trees on their own property. The National Arbor Day Foundation also encourages the planting of trees across America.

Hard-Working Volunteers

Another problem for animals is the decline of existing healthy habitats. In our modern world, with abundant pollution and with a great variety of introduced species of plants and animals, keeping a habitat in something approaching its original, healthy state can take a lot of work on the part of dedicated people.

In Marin County, California, volunteers for the Habitat Restoration Team work on weekends to remove non-native plants from the Golden Gate National Recreation Area, a spectacular stretch of hilly land along the entrance to San Francisco Bay. In the springtime, native plants like California poppies and blue and yellow lupines cloak the hills with color. But non-native plants that are well adapted to the cycle of wet and dry seasons typical of the California coast also thrive here. Scotch broom is perhaps the best example of a successful exotic plant. It forms dense thickets all over Marin County, choking out native vegetation. To keep such invaders in check, fifteen to twenty volunteers show up weekly for all-day work parties. "They've done tens of thousands of dollars of work that we could never have paid for," says John Martini, a ranger in the area. "What you can't put a dollar figure on is what they've done to restore the environment."[5]

The work team was formed in 1988 to help restore and preserve the area in its natural state. Their efforts have been so successful that they have become a model for similar efforts in other parks.

Cleaning Up the Beaches

The Marine Mammal Protection Act passed by Congress in 1972 protects the lives of animals such as seals and dolphins from hunters. But it cannot protect them from the kinds of pollution that can injure or kill them. One of the most dangerous substances to marine mammals is plastic. Animals get tangled up in plastic fishing nets and cannot swim. They put their heads through floating plastic six-pack rings, which become imbedded in their necks, cutting into their flesh. They suffocate when their heads get stuck in plastic bags. These tragedies add up. For example, each year around 30,000 northern fur seals die from becoming entangled in plastic, mostly in net fragments.[6]

Birds and mammals alike also eat plastic, mistaking it for food items. Bits of floating plastic can look like fish eggs and small crabs to birds and other animals. In Alaska, 40 percent of seabird species will eat plastic, and plastic has been found

Volunteers help clean up the shores of Lake Michigan as part of the
annual national beach cleanup.

in the stomachs of whales and dolphins. Sea turtles, all of which are either endangered or threatened, often swallow plastic bags, mistaking them for jellyfish.

Plastic and other marine debris can also endanger humans who use the oceans. Net fragments can become entangled in propellers, and plastic bags can block engine cooling intakes. At the end of 1988, a law forbidding dumping plastic trash into U.S. waters was passed, and limits were also set on other kinds of trash. But not everyone pays attention to the law.

It is hard to clean floating debris from the open waters of the ocean. Much of the debris gets washed up on beaches. The Center for Marine Conservation promotes beach cleanups around the world. Volunteers from Canada, Guatemala, Japan, Mexico, Taiwan, and the United States all take part. Much of the effort goes into the Fall Beach Cleanup, which generally takes place in September of every year. In 1990 more than 100,000 volunteers, mostly children, cleaned the shores of lakes, rivers, and streams as well as the East and West coasts.

9
National Wildlife
Refuges

The National Wildlife Refuge System, managed by the
U.S. Fish and Wildlife Service, is very special. It is the only
network of public lands in the world that is set aside
specifically for the conservation and improvement of wildlife habitats. Over 91 million acres of land in the United
States, from the far north of Alaska to the southern parts
of Florida, are protected by the more than 470 refuges
within the System, as are parts of the U.S. Virgin Islands,
American Samoa, and Puerto Rico. New refuges are regularly added, increasing the habitat saved from development
each year.

Refuges protect an enormous variety of habitats.
Along the coasts, marshes where fish and crustaceans
breed are protected, along with mudflats where shore-
birds feed. Southern swamps like Okefenokee in Georgia
provide homes for alligators, egrets, snapping turtles, and

an enormous variety of other animals. Eastern forests, a habitat that was reduced early on in the history of our nation as settlers felled trees to create farmland, are protected in refuges, as are western deserts and midwestern prairies. Such a variety of habitats provides homes for more than sixty endangered species, including whooping cranes, manatees, and bald eagles, as well as endangered plants. Hundreds of other animals, like sea turtles, elk, bison, bears, raccoons, and dozens of kinds of birds, live on refuges.

What Is a Refuge?

Each National Wildlife Refuge was set aside for a particular legal purpose. The very first refuge, Pelican Island in Florida, was established by President Theodore Roosevelt in 1903 to provide protected habitat for birds like egrets that were being mercilessly hunted for their feathers. Antioch Dunes National Wildlife Refuge in California exists to protect two endangered plants and an endangered butterfly. The purpose of a refuge can shift over time. Bosque del Apache National Wildlife Refuge along the Río Grande in New Mexico was originally established to protect the then-endangered greater sandhill crane. Today, the greater sandhill is no longer endangered, but the refuge is home to other endangered species such as the bald eagle and peregrine falcon. In addition, a few precious endangered whooping cranes make their winter home on the refuge.

For many years, a particular emphasis in the refuge system was providing habitats for migratory waterfowl—ducks, geese, and swans. These birds nest in the north and fly south for the winter. Waterfowl depend on wetlands—marshes, swamps, and coastal waters—for nesting, migration stopovers, and wintering grounds. The vast majority of refuge land—77 million acres—is in Alaska, where countless thousands of birds breed every year. A heavy concentration of refuges also exists in the prairie pothole region of the northern Midwest, especially in Minnesota and North Dakota, where large numbers of waterfowl breed. A series of National Wildlife Refuges lies along each of the nation's four major waterfowl migration routes—the Atlantic, Mississippi, Central, and Pacific flyways.

Much of the money used to buy and improve waterfowl habitat comes from the sale of Duck Stamps. When a waterfowl hunter buys a hunting license, he or she also must buy a Duck Stamp, which costs $15.00. The money from Duck Stamp sales—more than $20 million in 1990—goes toward buying and improving waterfowl habitats.[1]

Managing Refuges

Unlike some protected lands, such as wilderness areas, most refuges are actively managed to benefit wildlife. They are not preserved—left in a natural state—they are conserved—restored and enhanced in a variety of ways. The

Sandhill cranes feed on grain at Bosque del Apache National Wildlife Refuge in New Mexico.

land is often modified to increase its benefits for wildlife. For example, ponds and water channels are dug to provide open water for birds. Grain like wheat or corn may be grown and left in the fields as food for the birds. Some species, like the endangered Attwater prairie chicken and the sandhill crane, benefit from periodic burning of grasslands; controlled intentional burns on refuges for such species helps maintain the kind of habitat the animals need.

Many refuges, however, are left as much in their natural state as possible, since species such as the Kodiak bear in Alaska and the piping plover on the East Coast do best in an undisturbed habitat. Some refuges, or parts of refuges, are also wilderness areas where the habitat remains natural. No roads are built, and people can only enter on foot or on horseback.

Over the years, the refuge concept expanded to include an effort to include as great a variety of American habitats as possible in the National Wildlife Refuge System. As a result, refuges today represent every type of habitat, from Hawaiian rain forests where it rains almost every day to Arizona deserts where only an inch of rain may fall each year. Altitude and temperature vary enormously as well—Desert National Wildlife Range in Nevada features 9,000-foot mountains, while the Salton Sea National Wildlife Refuge in California lies 226 feet below sea level; temperatures at Havasu National Wildlife Refuge in Arizona regularly soar over 38°C (100°F), while the wind chill at Des Lacs National Wildlife Refuge in North Dakota can plunge to minus 85°C (minus 105°F).

In recent years, the National Wildlife Refuge System has placed greater emphasis on managing refuges to benefit all the native species present rather than just the waterfowl or the endangered species. Preserving biological diversity can be difficult, as actions that favor one sort of native plant or animal can cause problems for other species. Managing for diversity is a juggling act. But in the long run, it helps all native wildlife. A healthy habitat makes a good place for all the resident species to thrive.

People and Refuges

Many wildlife refuges also provide opportunities for people to enjoy nature. They have nature trails and auto loops for viewing the plants and animals that live on the refuges. Wildlife refuges are great places for bird-watching. Education is also an important purpose for wildlife refuges. Many have visitors' centers with displays explaining the purposes of the refuge and describing the habitats, plants, and animals that can be seen there. Refuge personnel and volunteers give talks and lead nature walks, pointing out interesting features of the area.

On many wildlife refuges, volunteers are especially important to successful operation. Throughout the U.S. Fish and Wildlife Service (which includes fish hatcheries, research stations, and offices as well as refuges), volunteers put in over a million hours in 1990.[2] There are no age restrictions for volunteers.

This bridge at the San Francisco Bay National Wildlife Refuge gives visitors a way to explore and view wildlife.

What do volunteers do? At the San Francisco Bay National Wildlife Refuge, for example, volunteers, who can choose the hours they wish to work, lead nature walks, provide information at the visitor center, pick up litter, install signs, and so forth. The refuge gives a twenty-hour training course three times a year for volunteers. In 1990 trained volunteers donated 18,000 hours, as much time as would be put in by nine paid employees!

Kilauea Point National Wildlife Refuge in Hawaii relies very heavily on volunteers. The refuge has 125 volunteers and only six employees. Since 400,000 visitors pass through each year, there is plenty to keep everyone busy. In addition to helping on the refuge by leading hikes, running the gift shop, and performing other jobs, the volunteers help with a newsletter distributed four times a year to every fifth grader on the island of Kauai. On an Oregon refuge, young people helped restore native vegetation to a slope that had been ripped up by off-road vehicles.

Refuges welcome groups such as scout troops that want to perform one-time projects like replanting native vegetation in a damaged area or constructing a new nature trail. Since many refuges border on waterways, the national beach cleanup helps make them safe and appealing places for wildlife and humans alike. At San Francisco Bay National Wildlife Refuge, one-time volunteers gave 14,000 hours of their time in 1990 to improve and maintain the quality of the refuge.

10
Reaching Out to the World

It is easiest to see changes in habitats, for better or worse, in our own backyards, neighborhoods, school yards, and parks. When we see migrating ducks and geese fly overhead, we can imagine them resting and feeding at wildlife refuges across the country. When we are lucky, we get to travel to national parks and wildlife refuges on vacation to experience nature and observe wild plants and animals in their own environments. But few of us have the opportunity to travel to distant lands, to see in person the wildlife portrayed so vividly on television specials. We have a much better chance of visiting a zoo, where samples of the incredible variety of life on our planet are exhibited. Even when we see the animal as a zoo captive, experiencing it in person is far more powerful than seeing it on television.

For decades, zoos were mainly places where animals captured in the wild were put on display for people to see.

Zoos sent collectors to gather specimens and tried to obtain as great a variety of animals as they could. The animals were usually kept in small, barren cages with bars, like prisoners in jails. Attention was paid to the physical health of the animals—the concrete floors and bare surroundings made it easy to keep the cages as antiseptic and free of parasites as possible.

But animals have important needs beyond the physical. When I visited the zoo as a child, I wanted to cry for the African hunting dogs that paced back and forth in their pitifully tiny cage. I knew they belonged on the African plains, where they could run for miles without encountering a fence. I saw a huge male gorilla in the ape house, sitting frozen in place with glazed eyes, all alone in a barren cage. He appeared to have gone mad with loneliness and boredom.

The Modern Zoo

Fortunately zoos are rapidly changing their goals. As different species of animals became rare in the wild, zoos became concerned about breeding them in captivity so they could provide themselves and other zoos with specimens. But many species wouldn't breed in the sterile zoo environment. The desire to breed the animals led to concern for the animals' psychological needs as well as their physical ones. Primates, for example, were given structures to climb on and toys to play with. Zoos were rewarded by an increase in breeding among captive species.

Gorillas at the Atlanta zoo are provided with a relatively natural habitat, allowing them to avoid boredom.

Visitors also enjoyed seeing active animals, and people concerned with animal welfare put pressure on zoos to provide animals with surroundings more like their wild homes. Today, many zoos feature naturalistic exhibits in which animals live with plants that would be familiar to them in the wild. Such exhibits give zoos the opportunity to educate visitors about the importance of habitat and to demonstrate that animals cannot be separated from the habitats in which they are found. They also provide animals with a much more interesting environment than the old bare cages.

Unfortunately, naturalistic exhibits sometimes are not as satisfying for the animals. It costs a lot of money to put together these exhibits, money that might have gone to preserve more land for animals. Inexpensive objects such as old tires, burlap sacks, and empty beer kegs make great "toys" for animals like bears and primates, but they don't fit in with a natural habitat. In the zoo environment, where animals are kept in cramped quarters, the plants sometimes must be protected from animals that might cause damage by climbing on them or eating them, so some zoos electrically charge the plants. If an animal touches the plant, it gets a shock, so it quickly learns to leave the plant alone. The zoo visitor sees what looks like a natural environment. But in fact, no captive home can come close to duplicating the wild.

Zoos and Wild Habitats

Thanks to Norman Gershenz, a keeper at the San Francisco Zoo, and Leslie Saul, Insect Zoo director, zoos are also becoming aware of their debt to nature. After all, if there were no wild animals in wild places, zoos would not exist. In the late 1980s, Gershenz and Saul met Dr. Daniel Janzen. Dr. Janzen has devoted his life to preserving the Guanacaste National Park in Costa Rica. Janzen's comment that zoos didn't do anything for conservation got the pair thinking. They realized that zoos were concentrating on saving individual species one at a time and were ignoring the importance of their habitats. When they came back to San Francisco in 1987, Gershenz began to think about how zoos could help preserve the wild habitats that provided them with such a wonderful variety of animals. The result was the Ecosystem Survival Plan™, co-founded with Saul. The program is sponsored by the American Association of Zoo Keepers, Inc., and hosted by the San Francisco Zoological Gardens. A key element in the plan is the Conservation Parking Meter™. A person puts in a coin, turns the knob, and a bright blue hummingbird flies across the face of the meter. With that act, one more tiny bit of wildlife habitat has been preserved.

"I began to wonder how zoos could help make a change," Gershenz explained. I started playing with the idea of change—how small change, individual nickels, dimes, and quarters, can make [a] change. That's when I first came up with 'Give your change to make a change.'

A hundred and thirty million people visit zoos every year. If we could get a quarter from each of them, that would be about 32 million dollars a year specifically for the preservation of threatened and endangered habitats.

"But how to do it? I thought and thought about it—what is familiar to everyone in America? And it came to me—a parking meter! A parking meter takes change. It's indestructible. We could keep the theme of conservation by recycling old parking meters, placing them in zoos across the United States. When it says 'Twenty-five cents saves ninety square feet,' young people can take a great deal of pride knowing they've done something."[1]

The parking meters have become a big hit. The second meter has been installed at the National Aquarium in Baltimore, and about twenty more are going in at zoos, aquariums, and botanical gardens across the country. Gershenz has also gotten calls from banks, Rotary clubs, libraries, high schools, and other organizations and businesses interested in the meters.

The parking meter project shows the power of an idea—how creative individuals can set a process in motion that makes an important difference in the world.

Adopt an Acre

The other part of the Ecosystem Survival Plan™ is the Adopt an Acre™ program. By contributing money, individuals or organizations can "adopt" land in a national park or preserve in a country like Costa Rica. The money

Norman Gershenz stands with the Conservation Parking Meter he helped design.

collected through the parking meters and the Adopt an Acre™ program is used to buy and preserve critical wildlife habitats. The protected areas are set aside as parks.

The parks are chosen carefully. The Ecosystem Survival Plan works in partnership with The Nature Conservancy. The Nature Conservancy has been in the business of purchasing land to be protected from development for many years. It has the resources to pinpoint habitats that are critically endangered and biologically significant. Experts from The Nature Conservancy study the plants, animals, and people in an area to see what can be done to preserve the entire system. Local people are a critical element—they have to be behind conservation, or it will not work. Conservancy staff members help train people in the country to organize and run nonprofit organizations that buy land and work to preserve the parks.

When Gershenz chooses a park from those The Nature Conservancy recommends, he takes other things into account, too. "Scientific research—I'm interested in seeing that occur on the site if possible. It should be visible—the people that live in the country should be able to utilize this area that's been saved by people around the world. And when young people in the United States today are of an age to travel to these places, they can do it. It's a partnership they've created. Young people can create a bridge to the wild for themselves. When they get to the other side, there will be this incredibly diverse fantastic living library that they'll be able to explore. Maybe some of them will become scientists, or maybe they'll just become people who enjoy nature and are so proud of what they've helped accomplish. They'll feel the pride of participation in these endangered spots around the world."[2]

In addition to working through the Ecosystem Survival Plan, The Nature Conservancy administers its own Adopt an Acre™ program.

Worldwide Preservation

Around the world, governments and private conservation groups are setting aside habitats as parks and preserves. But habitat preservation is a race with time, for wildlands are rapidly disappearing before the chain saw and the plow everywhere. The contrast between preservation and destruction is nowhere more clear than in Costa Rica. This Central American country is about the size of New Hampshire and Vermont put together. Unlike most of its neighbors, it has had a stable democratic government for many years and has no army. Its national parks are a model for other countries and cover about 12 percent of its land.

Unfortunately, the deforestation rate in Costa Rica is also high—each year from 1973 to 1989, about 2.6 percent of the country's forests disappeared, a higher rate than most nations.[3] The facts are bleak: "In 1950, 72 percent of Costa Rica was covered in forest. In 1973, it was 49 percent; in 1978, 34 percent; and in 1985, 26 percent. At this rate, Costa Rica's unprotected forests will be destroyed by the year 2000."⁴ The causes are the same as elsewhere—an increasing population, a thriving timber industry, expanding cattle ranching.

A variety of forces are combining to help save Costa Rica's natural resources. Efforts like the Adopt An Acre program provide valuable money to purchase parklands. Conservation groups such as The Nature Conservancy are cooperating with aid organizations like CARE (Cooperative for American Relief to Everywhere) to combine environmental preservation with education. The Children's Rainforest project is also saving important forest lands in Costa Rica. Private preserves like Monte Verde, which protects the precious cloud forest, and Rara Avis, a rain forest preserve where tourism is combined with programs to use the forest without destroying it, are showing that conservation can be economically profitable.

Every Bit Helps

Bit by bit, little by little—that's the key to saving wildlife habitats around the world, to making it possible for elephants, jaguars, dung beetles, orchids, parrots, ants, mahogany trees, toads, and all the other wonderful living things in our world to stay wild, to have secure homes so they can live out their lives naturally.

These habitats also provide homes for the human imagination and the human spirit. Just knowing wild places exist, that one day we may be able to visit them and witness the incredible variety and beauty of nature on our planet is reason enough to devote some of our time, effort, and money to this task. When wilderness disappears, something within us dies, too. We need wild places, and they need us—to help protect and preserve them. The pages that follow provide ideas for things you can do to get involved in saving our wild places.

What You Can Do

- Give a school report on loss of wildlife habitats. You could focus on your town, your state, the country, or the world. Contact government officials to find out what is being done in your area, and use your public library as a resource for material. By reviewing recent issues of conservation-oriented magazines, such as *Buzzworm, Wildlife Conservation, National Wildlife,* and *International Wildlife,* you can find out the latest information in this rapidly changing area.

- Suggest taking on a habitat preservation project to your school class or an organization to which you belong, such as a scout troup. You could inquire at a nearby wildlife refuge about possible projects, or you could raise money for habitat preservation and donate it to a program such as the Adopt an Acre™ program.

- Set up a bird feeder at your home. Keep it supplied so birds can count on it throughout the year for food. Try different seed mixes and provide suet for the birds during the wintertime. Using a guidebook to identify the birds, keep a log of which species you see at different times of the year.

- Plant native plants and flowers in your garden for wildlife, and supply a constant water source. For help, you can call the National Wildlife Federation at 1-800-432-6564 and order the Backyard Wildlife Habitat information packet, #79919 ($4.95), or the Gardening with Wildlife kit ($29.95).

- Contact the U.S. Fish and Wildlife Service, (202)208-5634, 18th and C Streets NW, Washington, DC 20240, about volunteering on a wildlife refuge in your area, or contact the refuge directly.

- Subscribe to a magazine such as *Buzzworm* or *With the Grain* so you can keep in touch with environmental issues. Better yet, get a classroom subscription.

- Contact the Council for Solid Waste Solutions, (202)371-5319, 1275 K Street NW, Suite 400, Washington, DC 20005, for their guide, *How to Set up a School Recycling Program*, and get your school involved.

Chapter Notes

Chapter 1

1. Janine M. Benyus, *The Field Guide to Wildlife Habitats of the Eastern/Western United States* (New York: Fireside, 1989), for information on what plants and animals live in different habitats.

2. Jim Watson, "The Last Stand for Old Growth?" *National Wildlife* (December-January 1990), pp. 24–25.

3. Jack Connor, "Empty Skies," *Harrowsmith* (July/August 1988), pp. 35–37.

4. Dr. Joseph Ball, Montana Cooperative Wildlife Reserve Unit, University of Montana, phone interview with author, Sept. 23, 1991.

Chapter 2

1. William A. Niering, *Wetlands* (New York: Chanticleer, 1990), p. 62.

99

Chapter 3

1. Jud Moore, U.S. Forest Service, Missoula, Montana, personal letter to author, Sept. 20, 1991.

2. Paul C. Pritchard, "The Best Idea America Ever Had," *National Geographic* (August 1991), p. 36.

3. Linda Wood, The Nature Conservancy, Helena, Montana, personal letter to author, Sept. 17, 1991.

Chapter 4

1. Kathryn S. Fuller, "How Big Is Your Backyard?" *Focus* (March/April 1991), p. 2.

2. The Earthworks Group, *50 Simple Things You Can Do to Save the Earth* (Berkeley, Calif.: Earthworks Press, 1989), p. 36.

3. Ibid., p. 13.

4. Paul Ehrlich, *Healing the Planet* (Reading, Mass.: Addison-Wesley, 1991), p. 8.

5. Kids for Saving Earth, fact sheet.

Chapter 5

1. Craig Tufts, *The Backyard Naturalist* (Washington, D.C.: National Wildlife Federation, 1988), pp. 20–28.

2. Ibid., pp. 64, 66.

3. Ibid., p. 65.

4. Ibid., p. 65.

5. Ibid., p. 41.

6. Ibid., p. 36.

7. National Wildlife Federation Backyard Habitat Program, Dept. N, 1400 16th St. NW, Washington, D.C. 20036-2266.

Chapter 6

1. Neil Andre, phone interview with author, Sept. 29, 1991.

2. Judith Southworth, "Operation Green Thumb," *Free Spirit* (Spring/Summer 1989).

3. Roger Cohn, "Calabaza Pumpkins Are Growing in the Bronx!" *Audubon Magazine* (July/August 1991), pp. 77–88.

4. Ibid., p. 81.

Chapter 7

1. Bob Jones, science teacher at Mead Junior High School, Spokane, Washington, personal letters to author, Oct. 20, 1991.

2. David Seideman, "Wading into the Fight," *National Wildlife* (December/January 1991), pp. 34–37.

3. William Hammond, phone interview with author, Oct. 15, 1991.

Chapter 8

1. Peter Steinhart, "No Net Loss," *Audubon* (July 1990), p. 18.

2. U.S. Department of the Interior, Fish and Wildlife Service, "50 Years Restoring America's Wildlife, 1937–1987," pamphlet.

3. John P. Wiley, Jr., "Phenomena, comment and notes," *Smithsonian* (March 1989).

4. Global ReLeaf, fact sheet.

5. Maura Thurman, "Striking Back at Marin Invaders," *Marin Independent Journal* (September 3, 1991), p. 81.

6. Center for Marine Conservation, "Marine Debris Facts and Figures," fact sheet, no date known.

Chapter 9

1. Nancy Marx, phone interview with author, Oct. 15, 1991.

2. Sheila M. McCartan, Interpretive Specialist, San Francisco Bay National Wildlife Refuge, personal interview with author, Sept. 20, 1991.

Chapter 10

1. Norman Gershenz, personal interview with author, Sept. 19, 1991.

2. Ibid.

3. Lisa Jones, "Central America," *Buzzworm: The Environmental Journal* (September/October 1991), pp. 71, 78.

4. Beatrice Blake and Anne Becher, *The New Key to Costa Rica* (San Jose, Costa Rica: Publications in English, 1990), p. 33.

Where to Write

Center for Marine
Conservation
National Beach Cleanup
1725 DeSales St. NW,
Suite 500
Washington, DC 20036

Clinton Hill's Kids for
Saving Earth
620 Mendelssohn,
Suite 145
Golden Valley, MN 55427

The Dolphin Defenders
812 N. Union St.
St. Louis, MO 63108

Ecosystem Survival Plan
San Francisco Zoo
Adopt an Acre
1 Zoo Road
San Francisco, CA 94132

Global ReLeaf
American Forestry
Association
P.O. Box 2000
Washington, DC 20013

The Green Guerillas
625 Broadway
New York, NY 10012

National Arbor Day
Foundation
100 Arbor Ave.
Nebraska City, NE 68410

National Institute for
Urban Wildlife
10921 Trotting Ridge Way
Columbia, MD 21044

The Children's Rainforest
P.O. Box 936
Lewiston, ME 04240

The National Wildlife
Federation
1400 16th Street NW
Washington, D.C.
20036-2266

The Nature Conservancy
1815 N. Lynn St.
Arlington, VA 22209

The Society for
Ecological Restoration
1207 Seminole Highway
Madison, WI 53711

FURTHER READING

Earth Beat Press. *Good Planets Are Hard to Find.* Vancouver, British Columbia Canada: Earth Beat Press, 1989. A booklet on ecology for children ages 8 to 14 and adults, subtitles "An environmental information guide, dictionary, and action book for kids (and adults)," which shows how to become actively involved in helping the environment.

EarthWorks Group, The. *50 Simple Things Kids Can Do to Save the Earth.* Kansas City, MO: Andrews and Mcmeel, 1990. Similar to the following book, but oriented especially towards children; includes explanations of some environmental issues.

EarthWorks Group, The. *50 Simple Things You Can Do to Save the Earth* Berkeley, Calif.: Earthworks Press, 1989. Guidelines on how to change your life to help our planet.

Erlich, Paul, *Healing the Planet.* Reading, Mass.: Addison-Wesley, 1991. A complete reference on today's environmental problems and how we might solve them.

MacEachern, Diane, *Save Our Planet*. New York: Dell, 1990. The subtitle, "750 Everyday Ways You Can Help Clean Up the Earth," explains the contents of this volume.

Pringle, Laurence, *Living Treasure: Saving Earth's Threatened Biodiversity*. New York: Morrow, 1991. An attractive book that discusses the diversity of life on earth and what can be done to save it.

Pringle, Laurence, *Restoring Our Earth*. Hillside, NJ: Enslow, 1985. Discusses restoring different types of habitats.

Stokes, Donald and Lillian, *The Bird Feeder Book*. Boston: Little Brown, 1987. Describes how to set up bird feeders, describes bird behavior, and has color photos and descriptions of common feeder birds.

Tekulsky, Matthew, *The Hummingbird Garden*. New York: Crown, 1990. A guide on how to attract hummingbirds to your home.

Xerces Society, The, *Butterfly Gardening*. San Francisco, CA: Sierra Club Books, 1990. Tells how to grow the kinds of flowers that will attract butterflies to your garden.

Magazines

Audubon, National Audubon Society, P.O. Box 52529, Bouldern CO 80322. Often contains articles relating to habitat perservation.

Buzzworm: The Environmental Journal, P.O. Box 6852, Syracuse, NY 13217. An independent publication filled with information about habitat perservation and other environmental issues.

National Wildlife and International Wildlife, National Wildlife Federation, 1400 Sixteenth Street NW, Washington, DC 20036-2266. Often have articles relating to habitat perservation.

With the Grain, With the Grain Foundation, P.O. Box 517, Mattawan, MI 49071-0517. Devoted to educating people about their effects on the environment; accepts no advertising.

Index